Leonardo's Machines

*Insights into Leonardo da Vinci's engineering notions
with four paper models to cut out and glue together*

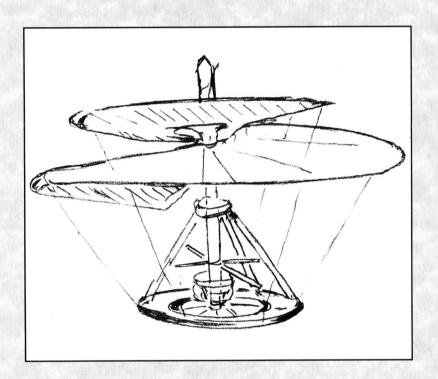

Bernard Ambrose

Leonardo's Machines

Introduction

This book with its four models to make is a tribute to one of the most prolific and inventive people ever to have lived. He painted, he sculpted, he designed, he drew, he invented, he studied, he observed and above all he applied himself in a creative way to the problems of the day. More than anyone else he was the inspiration for the idea of 'Renaissance Man', a creative generalist, skilled and cultured in many diverse fields and exultantly different from the narrow specialists of today. During his lifetime he filled dozens of notebooks with ideas and studies and fortunately many have survived. He drew on sheets of different sizes, never very tidily and with each page seemingly unrelated to others. After his death these sheets were gathered together into collections and were often glued to backing sheets. The two major collections are called the 'Codex Atlanticus' which is in the Biblioteca Ambrosiana in Milan and the 'Codex Madrid' which is in the Biblioteca Nacional in Madrid. There are at least 21 further collections of his works and all have been produced as facsimiles by Giunti of Florence, offering lifetimes of study to those who are interested in investigating his ideas.

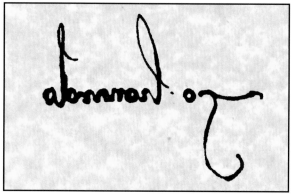

Leonardo's Signature

Leonardo was left-handed and it is interesting to note that most of his writing was done from right to left across the page in mirror writing. Some people have said that he wrote like this in order to keep his ideas a secret but it seems more likely that he found this style easier because of his left-handedness. Whatever the reason, it makes the manuscripts rather harder to read and introduces more difficulty into interpreting the meaning in his commentaries.

In this book we focus on Leonardo the engineer. His great contribution was to see machines not as complete in themselves but as made up of parts or 'organs'. It makes sense to see his sketches and diagrams as a kind of dictionary or encyclopedia of these parts and how they might be linked together. Each of these 'elements of machines' had a single function but they could be arranged in a virtually infinite number of ways to produce machines which would perform any task that the human imagination might demand.

This idea is so completely accepted nowadays that it is very hard for us to imagine a time before this had happened. What Leonardo did not realise or perhaps he did but appreciated that it would require too much effort for a single lifetime, was that machines also need testing and refining if they are to work satisfactorily. Engineering ideas start with a simple sketch on paper and progress to the building of a prototype model. It is at this stage that flaws in the original idea come to light and can be corrected on the plan before proceeding with a rebuild. What we nowadays call 'Research & Development' or 'R & D'.

His own imagination soared to produce ideas and machines that seemed to foresee the future but they often included mistakes and misunderstandings which meant that they would never have worked. Very few of Leonardo's drawings show modifications to the original idea. However, even these deficiencies are interesting because they draw attention to the nature of invention and make us realise the importance of things that are so often taken for granted.

Leonardo's Machines

A Short Personal History

Leonardo da Vinci was born on the 15th April, 1452 in the hill town of Vinci in Northern Italy, half-way between the cities of Florence and Pisa. His father, Ser Piero da Vinci, was a lawyer and his mother Caterina a peasant girl.

Not very much is known about his very early life except that he only had the normal period of time at school and did not go on to University. His artistic and creative talents however were demonstrated at an early age when his father, having been asked to take a wooden shield to Florence to have it decorated, decided to let his son carry out the job. The young Leonardo set to work with a will to do something special and decorated it with pictures representing a monster emerging from a cave. When it was ready he set it up in his room, arranged the lighting to create a dramatic effect and invited his father in to see the finished result. Ser Piero was startled and amazed at what his son had produced and realised that he had a remarkable creative ability.

When Leonardo was 17 years old his father placed him as an apprentice in the workshop of Andrea Verrocchio in Florence. At that time such a workshop was engaged in a variety of activities such as architecture, sculpture, painting, etc. The work was commissioned by the nobility and the Church and would have given a young, inquisitive person like Leonardo a thorough insight into many techniques and processes. It was at this time that he was asked by Verrocchio to paint the head of an angel as part of a large picture showing the Baptism of Christ. It is said that when Verrocchio saw the result of his apprentice's work he resolved never to touch a paint brush again because he would never be able to equal the young man's skill.

Leonardo finished his apprenticeship and continued for some time in partnership with his master. He worked on several commissions but really he wanted to strike out on his own. He had heard that the Duke of Milan, Lodovico Sforza had an opening for a military engineer whose job it would be to advise the Duke on armaments and defences, so he composed a letter of introduction which described how he had the skill to improve existing weapons and to devise new ones. He also knew that Lodovico would be interested in his idea for a huge monument to commemorate his family so he included this in his letter.

Leonardo was not immediately accepted to the position but finally in 1482 he moved to Milan to start work. This was a challenging time for him. He really had no experience as a military engineer and some of his ideas were outrageous, but he persevered and excelled at whatever he did. He designed weapons but also painted portraits, staged pageants and planned for the construction of the monument for Sforza. This was to have been a huge bronze statue of Ludovico's father mounted on a horse and was to be the largest bronze casting that had ever been attempted. Preparations were well advanced when disaster struck. Milan was invaded by the French and the bronze for the monument was taken and used to make cannon. This was to no avail as the City was decisively defeated. It is not known how well his military ideas worked but Leonardo fled from Milan to seek his fortune elsewhere.

He visited the cities of Mantua and Venice before eventually returning to Florence. Things did not go very well for him but then Florence was attacked by forces led by the prince of Romagna, Cesare Borgia whom Leonardo had met briefly during his time in Milan. Cesare Borgia had been impressed by Leonardo's skill and invited him to become a military engineer for him. In this role, Leonardo drew up plans and advised on defensive structures. However, before long he became so upset by the treachery and dishonesty of the Borgias that he resigned the post and returned to Florence. One of the best known paintings in the world, the Mona Lisa, now in the Louvre in Paris, was painted at about this time. Although he was a supremely gifted artist he in fact completed very few pictures. He was never satisfied to use existing techniques and was always experimenting, trying out new ideas and using new materials.

His interest in how things worked drove him forward. He seemed to have had exceptionally acute eyesight and drew and described elements of the flight of birds, no doubt searching for a method of allowing humans to fly. He returned to Milan, then went back to Florence and then once more to Milan where he took a particular interest in anatomy by dissecting several corpses and drawing what he saw. He also studied geology and wondered how fossil shell-fish could have arrived high in the mountains. He was interested in air and water currents, botany, geometry, etc. and also speculated on how the world might reach its end in catastrophic turmoil.

His interest in geology shows in two very similar pictures which he made called 'The Virgin of the Rocks'. These show the Virgin Mary with Jesus and John the Baptist with an angel set in a cave-like environment. One of the pictures is now in the National Gallery, London and the other in The Louvre, Paris. In addition he wrote a treatise on painting which explained how an artist can use techniques to represent distance and how the tones and shadows used in a picture were important.

In 1516, Francis I, King of France invited him to live in the Loire valley at Amboise where he had a residence. He was then 64 and his health was failing. His sight was weak and his right hand became paralysed but he was an entertaining companion to the King and one can imagine the conversations they had discussing far reaching ideas.

The end came on the second of May 1519 when Leonardo died in peace.

A Misunderstanding

Because Leonardo was so gifted, prolific and inventive, many people fall into the trap of thinking that there were no engineers and machinery designers until Leonardo came on the scene. This idea is totally mistaken as people have been developing technology since the dawn of history. Ideas have been put into practice and lessons learned from success and failure. People such as Archimedes (287-212BC) had devised war machines which used levers to defend against invading ships. Nearer to Leonardo's time Mariano di Iacopo (1385-1458), known as Taccola, and Francesco di Giorgio (1439-1501) had worked on bold constructional projects such as a system for providing fresh water to the city of Siena, and Filippo Brunelleschi (1377-1446) had made elaborate cranes to help with the building of the dome on Florence Cathedral. Leonardo was observant and inquisitive and would certainly have been aware of what others were doing.

Some of his drawings are certainly amongst the earliest representations of many engineering ideas but this does not necessarily mean that he was their inventor. He may simply have been using his outstanding ability to draw in order to record a clever idea which interested him and which he had seen in a workshop or on a building site. On the other hand, his fertile brain must surely have produced many original ideas and improvements. It will probably never be known exactly which they were, but on balance it is best to give him the benefit of the doubt.

One cannot help but speculate what he might have achieved today with access to precision instruments and modern machine tools.

Leonardo's Machines

Leonardo as an Engineer

Let us start with Leonardo's ratchet winch from the Codex Atlanticus. It is not clear whether this is an original Leonardo invention or a drawing of a machine that he had seen awaiting repair in someone's workshop. However, we can certainly agree that it is an inspired piece of work and seems far ahead of its time. It also illustrates the amount of detail that is present in the drawings if they are examined carefully and by someone with a knowledge and interest in engineering techniques.

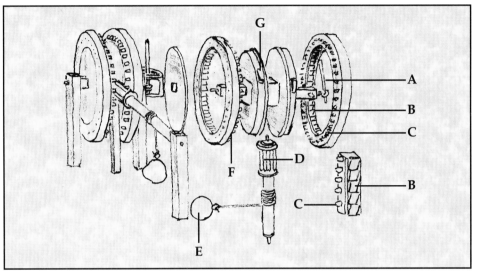

Ratchet Winch

A winch is a device for lifting heavy weights and the addition of the ratchet is of great value. It acts as a stop mechanism and prevents the load from slipping back. This is very useful because the operator can take a rest after a few strokes and not lose the height already gained.

The drawing consists of a picture of the winch fully assembled and also on the right-hand side an exploded diagram of the components needed to make it. Notice how he has moved the shaft and the load out of the way and rotated it by 90° to allow a clearer view. It is undoubtedly a drawing by an engineer and it can be studied to understand the method of operation.

Imagine that the handle (A) on the far right is pushed into the plane of the paper. The spring loaded pawl sticking out of the right-hand disc then engages with the ratchet (B) inside the right-hand ring and pushes it round. The teeth (C) on this ring then mesh with the cage-like gear (D) which is called a lantern gear. It is so called because its cylindrical cage structure makes it look like an old fashioned lantern. This turns the shaft and lifts the load (E). At the same time the lantern gear engages with the teeth on the left-hand ring (F) and turns it so that its ratchet slides over the pawl sticking out of the left-hand disc (G). When the pawl clicks into position the load cannot descend.

Now, when the handle is pulled outwards from the plane of the paper the situation is reversed. The left-hand pawl engages and the load will again be raised. What a clever idea! No matter whether the handle is pushed or pulled the load is raised.

However, it does not seem that this machine would allow the load to be lowered gently down again. This might not have been important as perhaps the problem that he was trying to solve only required that the load or weight be raised carefully and efficiently using the power of the human arm. After all, it is easy enough to let the load fall to the ground if it does not damage easily!

A Problem with Friction

Friction becomes important whenever two surfaces rub together. Heat is generated, energy is wasted and extra force has to be applied in order to overcome its effect. It presents a serious problem that all engineers have to deal with and much skill and genius in machinery design is devoted to finding ways to minimise its effect. In situations where a piece of machinery is powered by human effort, clever design becomes of crucial importance. More force than the operator can apply may be required simply to make anything move at all. Of course, nowadays we have powerful engines that can apply a lot of force, but unless the problem of friction is dealt with successfully, so much heat can be generated that components may catch fire or even melt.

One simple way to reduce friction is to spread oil or grease between the surfaces. In Leonardo's day that mostly meant using animal fats or perhaps fish oil, a far cry from the carefully formulated lubricating mineral oils of today.

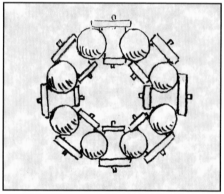

Ball Bearing

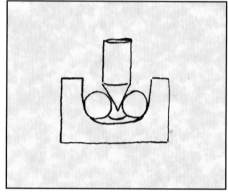

Thrust Bearing

In the Madrid Codex are two suggestions that Leonardo makes for dealing with the problem of friction. Both make use of the ball bearing. He realised that rolling metal balls have intrinsically less friction than flat surfaces, just as a rolling wheel has less friction than a sledge. A mechanism with ball bearings can be made even more efficient by lubricating it with modest amounts of oil or grease.

The first drawing is of a ring of balls held apart by a cage which prevents the balls from rubbing against each other. It is very similar to modern ball races. In those days the problem of manufacturing truly spherical balls was much greater than it is today. The modern method is to cast the steel into spherical shapes and then put them into a grinding mill where they are ground until perfectly spherical. Then they are polished and hardened. In Leonardo's day, the bearings would have been cast iron spheres similar to cannon balls and it was not possible to achieve such a high degree of surface smoothness. None the less, such a design would have been a great help in reducing the impediments to circular motion.

The other picture shows a bearing using three balls running in a cup and is designed to withstand the downwards thrust of a heavy rotating shaft such as that needed to support the centre post of a crane or a swing bridge.

It is curious that a device similar to Leonardo's three balls in a cup is used today to test the break-down qualities of lubricating oil. The rotating shaft is replaced by a fourth ball to which a load is a applied. As the speed of rotation and the load is increased a point will be reached at which the oil used to lubricate the balls fails. The information gathered from such a test enables one oil to be compared against another, ultimately leading to improved lubrication of moving parts.

Leonardo's Machines

Machines of War

In his position of military engineer, Leonardo drew many different ingenious machines. It is not clear how many of them were actually built but there are substantial advantages to be gained from inventing a clever new device both practical and psychological. There is the practical one in that it might work better than previous designs but there is also the psychological effect of surprise on the enemy. A cunning new machine might not work too well in itself but if it causes the enemy to hesitate and lose momentum that might well be sufficient to offer a decisive advantage to the home side. His ideas included various slingshots and catapults, devices for pushing invaders ladders away from a castle wall, transportable bridges for troops and even a man-powered military tank.

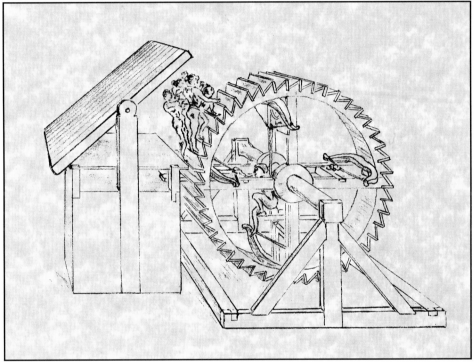

Rapid Fire Crossbow

It is clear that Leonardo very much appreciated the advantages of rapid fire in a battle and designed an ingenious crossbow which could be quickly reloaded. This design comes from the Codex Atlanticus.

The idea is to have a man sitting at the centre of a large treadwheel. His sole function is to prepare four crossbows for loading and to aim and fire them in turn. The energy in a crossbow is provided by tensioning a stiff bow. This is normally done by winding a screw thread, an effective but slow process. The classic battle of Agincourt in 1415 was won by the English with their longbows which although having less range than the French crossbows could be loaded and fired much more rapidly.

In Leonardo's design we see an alternative way to provide the necessary tension. Several men climb the steps of the treadwheel and force it to rotate. This rotation causes ropes fastened to the strings of the crossbows to wrap slowly around the stationary central axle of the wheel so pulling on the bowstrings and tensioning them. When one of the bows reaches the correct position for firing, someone pushes a plank through the steps to hold the wheel still and the crossbow can then be aimed and fired. In the drawing he fires towards the enemy on the the left. The strong timbers above the men climbing the wheel are to protect them from missiles or arrows being thrown or fired at them.

All seems to be in order and it appears to be an ingenious solution to an important problem. However, do look more closely at the drawing, in particular at crossbows two and three. When they move into the firing position they will be upside-down thus making the process of aiming more difficult. More importantly, the ropes tensioning the crossbow strings will actually unwind from the central shaft as the treadmill turns, so reducing the tension not increasing it. This is so surprising that surely it must be a deliberate mistake on the part of Leonardo to confuse anybody who tried to steal his idea! Surely it was!

He goes into considerable theoretical detail to work out the mechanics of the device.

Make the stem, which gathers the rope charging the crossbow, 1/3 thick, so that a turn gathers one ell (110cm) of rope. So, since the above named stem is one third, its half will be one sixth, and the lever is five ells, that is 30 sixths; and one sixth is the counter lever, so that you have 30 against 1. Therefore it appears clear that if you place over the head of the lever 20 men weighing 4,000 pounds, they will exert against the counter lever a force of 120,000 pounds, enough to charge the 4 crossbows.

What he is saying is that if the machine is friction free and the radius of the wheel is 30 times the radius of the centre shaft then the mechanical advantage of the lever is 30 and so the force exerted on the combination of the four ropes attached to the crossbows will be 30 times the weight of the men climbing the steps and that this will be sufficient to tension the bows.

What a formidable idea!

Attack and Defence

In Leonardo's time the best method of defence against an invading army was to retire inside some strong walls and to try to sit out the siege. The hope was that the defenders could last out long enough for the attackers to lose patience and leave and so the top priority was to have sufficient supplies of food and water within the castle. This was not quite so passive a situation as it sounds as it was also possible from time to time for the defenders to sally out of one of the gates and hope to catch the attackers unprepared and off-guard. Being outside, they of course were not protected by strong walls and so were easier to come to grips with. The time and the place of such a counter-attack could be chosen by the defenders.

In contrast, the attackers did not wish to lose the initiative or to have to wait for months for starvation to take its toll, all the time risking counter-attacks and the approach of a relieving army. Before the days of large cannon which could blast a hole in the walls, one method of attack was to use scaling ladders and to attempt to climb over. Even when a breach was made with a cannon or a catapult, an attack over the walls with such ladders at other places around the walls would serve to keep the defenders off-balance and prevent them from concentrating their forces around the breach.

For an attacker, if enough ladders could be placed around the walls at the same time then perhaps the defenders could be overwhelmed. For a defender, Leonardo's device for pushing scaling ladders away from a castle wall might be just enough to swing the balance in a critical situation. His design shows a thorough grasp of mechanical principles and it is interesting to examine it in detail.

An Interesting Solution

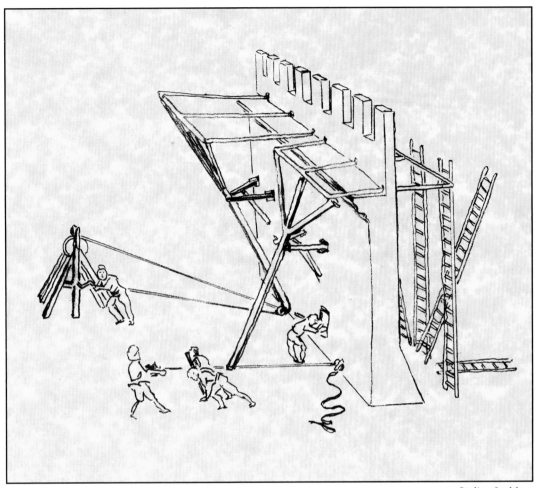

Scaling Ladders

Leonardo shows a contraption which three men could use to push the ladders away as long as the walls have been prepared first. To do this, a series of horizontal beams are placed along the walls close to the top and they are linked to secondary beams passing through the wall. These secondary beams act as pistons and are moved by means of levers. The idea is that when the attackers have placed their ladders against the wall the defenders pull on a rope fastened to the lever and this in turn pushes the pistons. The beams on the outside then push the ladders away from the wall and the attackers fall to their deaths.

Then the three men take up another rope to pull the mechanism back to its starting position ready for the next assault. It may be that such devices were already known and in use and Leonardo's contribution is a suggestion to improve their effectiveness. He had realised that if a winch is incorporated then it would only need one man to move the lever.

If we look carefully at his drawing we notice that the rope which returns the system to its starting position is connected to a larger diameter drum on the winch. This seems to be a sensible idea because it would require less force to pull the lever back than it would to push the ladders away. In addition, it would offer the advantage that the return movement would be quicker. There would therefore be less time for the attackers to get their ladders in position under the horizontal beams while they were extended away from the walls.

Thus the improvements offer both a reduction in manpower needed to operate the machine and a reduction in the machine's 'dead time', both typical Leonardo contributions.

A Steam Cannon

Explosives were being introduced during his lifetime although much fighting was still being done with spears, swords, bows and arrows. Cannons were effective but were seriously handicapped by the time and effort that they required to prepare them for another shot. Leonardo attempted to tackle this problem by showing how cannons with multiple barrels might be designed, a follow on from the rapid-fire crossbow described earlier. In this respect he perhaps anticipated the Gatling gun of the American West by more than two centuries.

Steam Operated Cannon

Leonardo was aware of the tremendous change in volume which occurs when water turns to steam and he wondered if this could be used as a substitute for the explosive charge in a cannon.

The idea was to heat one end of the barrel of a cannon into which a cannon ball was already loaded. A fire was made around one end of the barrel and when it was hot enough, a valve could be opened to let some water in. This water would be immediately turned into a great volume of steam and so eject the cannon ball!

Such a weapon might have made a significant difference if it had worked as planned. However, a modern engineer would have had considerable reservations about the effects of expansion on the barrel and shot as the heat was transmitted along it. He did foresee the future in the sense that such a steam operated cannon was built many centuries later. Unfortunately it was not a success. The closest modern machine to Leonardo's steam cannon is the one used on aircraft carriers to launch planes from the deck. However, there is a great difference in that the high pressure steam is first generated in a boiler and then only injected into the launch tube when it is needed.

A Question of Foresight

If only Leonardo had thought about replacing the cannon ball with a piston then engineering would certainly have developed at a different pace. The benefits of harnessing the power of steam were enormous and perhaps would have come rather earlier in the history of mankind. He never thought of using steam power for a vehicle as Richard Trevithick did in 1802, nor to operate machinery and pumps as was done with such success to initiate the industrial revolution.

Leonardo did foresee the Industrial Revolution in the sense that he appreciated the need for automating industrial processes and reducing the amount of skilled labour required to make things. This design for a file making machine gives a good insight into his way of thinking.

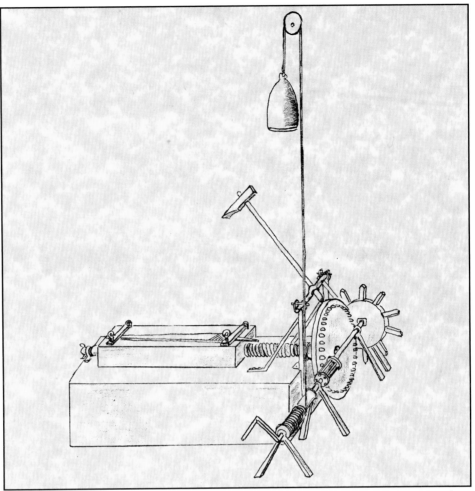

File Making Machine

At the time files were hand made from a steel blank. The teeth were raised, one at a time, by striking a chisel very hard with a hammer. Each tooth required a separate blow and in order to make an effective file capable of producing a smooth finish, the teeth had to be parallel. After the teeth were embossed into the steel, the file had to be hardened by heating and quenching as otherwise the metal would have been too soft to be used. This whole process required a very skilled craftsman and what is more, only one file could be worked on at a time. The advantages of automating part of such a process were considerable and once set up, the machine could be operated by unskilled labour. Indeed it might well become possible for that unskilled labourer to supervise several machines at the same time, thus greatly increasing the output of files and reducing the cost.

In the Codex Atlanticus there is a drawing which suggests combining the two tools, the chisel and the hammer into one by shaping the end of the hammer. The energy required for the manufacture is provided by the operator who winds a handle to lift a heavy weight to a high level. When this is released it turns a shaft on to which is fastened a spoked wheel or cam. The spokes catch on a lever which lifts the hammer and then releases it. Each time the hammer falls on to the file blank it raises a tooth and gearing moves the blank forward by the correct distance ready for the next blow to raise the next tooth. This coordination is achieved by a lantern gear engaging a crown wheel connected to a screw passing through a nut trapped under the block to which the file blank is fastened. As the screw turns so the file blank is moved forwards. The crown wheel is so named because its teeth stick out from its circumference like the spikes on a crown.

The principle of using a 'drop hammer' in this way was not new but the way of combining it with a self advancing screw device may be a real contribution by Leonardo. Of course, the next stage would be to connect it to a water wheel and so avoid the operator winding the weight upwards first. In that case it becomes much more possible to see how an operator could operate several machines at the same time.

However, there is a flaw in Leonardo's drawing which raises the question of whether this device was ever actually made and operated. If you were to build this machine as it is drawn, you would find that it would not work because there is a simple timing error. The hammer does not have sufficient time to fall to make its cut before the next spoke on the cam wheel comes along to lift it again. It requires some kind of governor or escapement to allow the hammer to do its work before this happens.

An Improvement to a Clock

The mechanisms used in clocks were of great interest to Leonardo. In his lifetime, sundials and astrolabes were commonly used for timekeeping and certain weight driven clocks had already been devised. Consideration was also being given to the use of a coiled spring to provide the power to drive clock mechanisms. If this could be achieved, then clocks could be made smaller and more compact, indeed they might even become portable! However, at the time no solution had been found to the problem that the spring exerts less force as it unwinds. This meant that spring driven clocks did not keep good time.

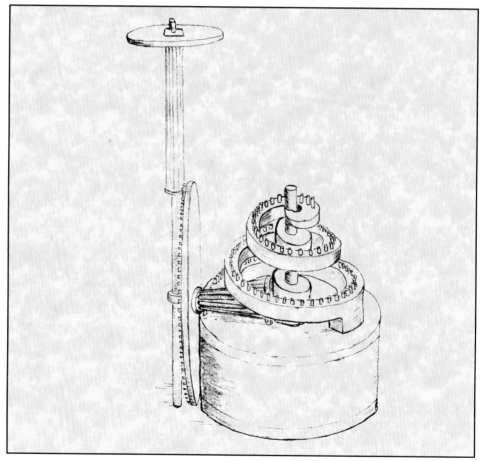

Volute

One of Leonardo's ingenious solutions is shown above. The clock spring is not visible as it is contained within the drum at the base of the mechanism.

When the spring unwinds it drives a curious spiral shaped gear mounted on a shaft. The teeth on this spiral engage with a cone shaped gear which turns the large crown wheel fastened to it. The teeth on the crown wheel engage with the elongated teeth on a pinion (a small diameter gear) which is shown on a vertical axis. Another gear wheel fastened to this pinion eventually drives the clock.

When the spring is fully wound the cone shaped gear will be at one end of the spiral and as the spring unwinds the conical gear will roll along to the other end of the spiral. Now you can see the need for the elongated teeth on the pinion which keep in contact with the teeth on the crown wheel as it moves up and down. When the conical gear is at the bottom of the spiral gear one rotation of the spring drum will turn the spiral gear once and this will turn the conical gear by many revolutions. However, when it is at the top of the spiral the conical gear will make fewer turns for one revolution of the spiral gear, so the effectiveness of the spring changes as it unwinds. It is important to realise that this mechanism is not a complete clock but merely a device to compensate for the smaller force exerted by the clock spring as it unwinds. It is not clear whether the spring is shown in the fully wound up or unwound position. I suspect it is shown in the wound up position. What do you think?

This device could have been used as a replacement for the more conventional fusee and drum in which a fine chain or a length of gut wound from a conical pulley to the spring drum in order to keep the driving force constant but it would have been difficult to make and trouble would have arisen if the cone shaped gear and crown wheel had jammed as they slid up the long pinion on the left. Here Leonardo was applying his thoughts to the solution of a very real problem of the day. Interestingly, it was not until Galileo (1564-1642) that the pendulum was thoroughly investigated and then became the basis of so many clocks.

Leonardo's Bicycle

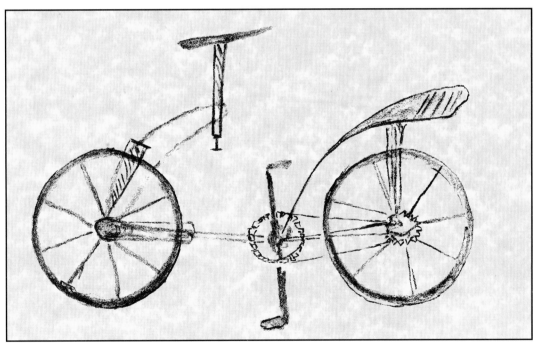

Bicycle

A drawing which has baffled many people since its discovery in 1974 is one of what to all intents and purposes is a modern bicycle. Is this yet another instance of Leonardo's genius as an inventor? At a stroke he appears to have invented both the idea of motion on two wheels and the idea of using a chain drive.

From the drawing of the bicycle it is not possible to see that he was suggesting the use of a metal chain but it is known that Leonardo was interested in chains and he had produced several possible designs for them. He had suggested that a metal chain would be better than a rope for lifting heavy weights but nowhere else does he suggest that it could be used for propulsion. How is it that the drawing of the bicycle was not discovered until 1974?

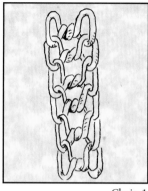
Chain 1

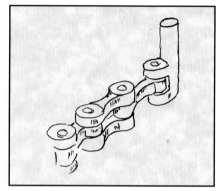
Chain 2

After Leonardo died, many of his papers were collected together and pasted on to large paper sheets known as Atlantic size sheets. That is why the Codex or book into which they were then bound is known as the Codex Atlanticus. This gathering together took place in the 16th Century and the bicycle drawing was discovered on the back of a sheet which had been folded in half and glued down at that time. During all the centuries that followed, nobody suspected that there was anything drawn within the double thickness.

With the passing of so much time, the Codex had fallen into a state of disrepair and a restoration was carried out in the 1960s by some Italian monks. In 1974, after the process was complete, a researcher named Hans Erhard Lessing was examining the repaired copy and discovered the bicycle drawing along with some other sketches. As might be expected, there was considerable excitement. It seemed that Leonardo had foreseen the invention of the bicycle. Even with its errors and imperfections, this would surely be yet another demonstration of his genius.

Because bicycles are common and relatively cheap nowadays and seem to be an example of simple and robust engineering, it is easy to lose sight of just how sophisticated a machine a bicycle really is. In Leonardo's time it was not known that it was possible to balance and ride along perched on only two wheels and there was the additional difficulty that a bicycle does need a relatively flat surface to ride on. People say that perhaps the reason why the Incas of South America, a very sophisticated civilisation in other respects, never invented the wheel was because their kingdom being so mountainous and steep had few places where a wheeled cart would be very useful.

In a similar way, it would have been very difficult to master a primitive bicycle on a rough, cobblestone road and it would have taken a considerable feat of imagination to realise that it could be done. If Leonardo had indeed managed to see through all these difficulties then perhaps this would have been his supreme invention.

Everyone agreed that the drawing was not in Leonardo's style but had it perhaps been done during his lifetime by one of his students and remained hidden all those years? There are other sketches on the same page which were similarly not by Leonardo but the name Salai is written nearby and this name belongs to one of Leonardo's young students. Could Salai have copied one of Leonardo's drawings and not done it very well? On closer examination of the bicycle drawing we can see that there are obvious errors with the way that the steering and pedals work but perhaps these are no more than you would expect if an inexperienced student had been trying to copy the picture.

The researcher who had found the drawing contacted Professor Carlo Pedretti, an expert on Leonardo's work and asked for his opinion. The Professor said he had held the folded sheet up to the light in 1961 and had seen the shadow of two circles within. However he had seen nothing more and certainly no bicycle! Furthermore, chemical analysis of the crayon used to make the drawing showed conclusively that it was a modern addition. Someone who had seen a modern bicycle or in particular an early bicycle of the Victorian era must have seen the two circles and then doodled over them. It seemed that the lines must have been added when the Codex was being restored. Leonardo's bicycle had taken everybody for a ride!

However, that is not the end of the story. The workers at the Restoration Laboratory at the Abbey of Grottaferrata, near Rome are offended by the assertion that they could be responsible for a deceit and deny that any forgery could had been done during their restoration.

This story indicates how careful we must be when trying to interpret documents of any kind and especially old ones like the Codex. It can be dangerous and misleading to take them at face value and also to jump to conclusions because they strike a chord with something we know about nowadays or would like to believe. However, we can be reasonably sure that Leonardo was not the inventor of the bicycle. That had to wait for a later generation.

Flying Machines and Ornithopters

Many people before Leonardo had dreamed of flying, and it was an idea that Leonardo found especially exciting. He observed the flight of birds and watched how they used their wings and tail to control their motion and to exploit currents of air hoping that he could discover how humans could fly using their own muscle power. The name 'ornithopter' is used for a flying machine based on the way birds and bats fly. Leonardo shows in great detail how one might research and construct such a machine and studied the best shape and covering for the wings. He concluded that they must be large and would probably be successful if covered with a membrane rather like the wings of a bat. He saw that there were many different shapes of wings in the animal kingdom and concluded that a wing testing machine would be the best way to discover and design the most efficient shape for human flight. One model in this collection illustrates how he thought that this should be done.

Two Possible Designs

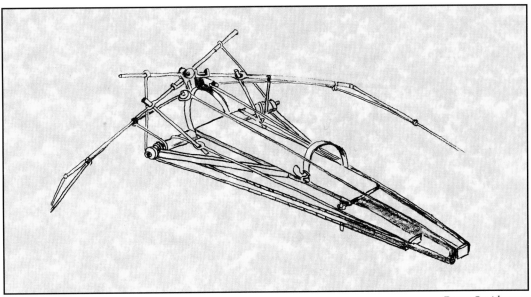

Prone Ornithopter

Leonardo's Machines

Leonardo soon realised that it would take a considerable expenditure of energy to be able to fly and that the required output could only be achieved if all muscles, especially the powerful ones of the back and legs could be brought into play. He started with a boat shaped flying machine in which only the arm muscles were used but he quickly abandoned that idea and then worked on ornithopters in which the person had to stand upright. This enabled his legs to work treadles and his arms to work levers at the same time. He even shows a rather uncomfortable design where the head presses against a beam to allow the back muscles to add to the action of the legs!

Another idea shows a machine suitable for a person in a prone position using mainly his leg muscles so that his arms and hands are free to control the stability of the ornithopter.

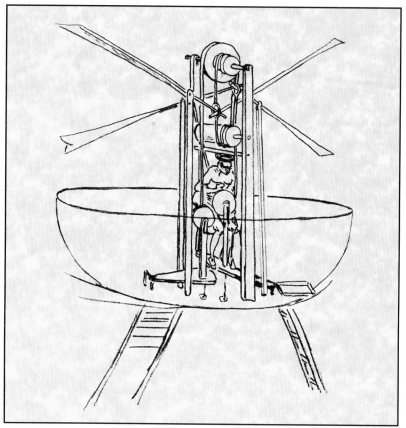

Standing Ornithopter

He also devised aeronautical instruments such as an anemometer to measure the wind speed and a hygrometer to measure the 'quality and thickness of the air', even a device for measuring the inclination of the flying machine was outlined. However, without a source of power other than that of a person's muscles it is sad to say that his ideas were to be doomed. It was not until the late twentieth century that mankind was able to fly using the power of human muscle. The first successful aircraft were called the Gossamer Condor (1977) and the Gossamer Albatross (1979) and the key to their success lay in the use of much lighter and stronger materials such as carbon fibre and aluminium that were never available to Leonardo.

As far as we know his ideas were not tried out at the time but he did write, *the great bird will take its first flight upon the back of the great swan, filling the whole world with amazement and filling all records with its fame; and it will bring eternal glory to the nest where it was born' and 'From the mountain which takes its name from the great bird, the famous bird will take its flight, which will fill the world with its great renown.* Did he manage to get an ornithopter built and launch it from a hill called the Great Swan? If only we knew.

Leonardo's Machines

A Description of the Models

We have chosen four of Leonardo's drawings to interpret as paper constructions and hope that you will enjoy making and displaying them. The models are not to any particular scale and we have had to simplify some of the parts in certain cases but they do give a feeling for Leonardo's intentions. A three-dimensional model is always more interesting to examine than a two-dimensional drawing.

Model 1. A Design for a Parachute

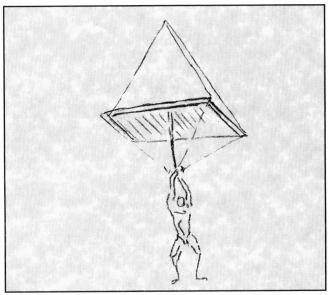

Parachute

Leonardo states, *If a man have a tent made of linen of which the apertures have all been stopped up, and it be twelve braccia* (about 24 feet or 7 metres) *across and twelve in depth, he will be able to throw himself down from any great height without sustaining any injury*. He envisaged the parachute to be a flying machine in its own right rather than as a means of escape from another flying machine in difficulty, its main purpose nowadays. He saw it as a machine which would allow a man to 'fly' down safely from a high rooftop or cliff. Perhaps he even envisaged the use of a brigade of Renaissance paratroops to astonish and terrify the enemy! His parachutist would have had to land quite close to the base but how excited he would have been to have seen our modern para-gliders. With these, people can launch themselves from the highest mountain peaks and steer themselves down to a safe landing quite far away. Such is the accuracy and control of modern machines that they can aim for and land on virtually any small flat area which is available. Such a happy outcome would not have been available to Leonardo's brave parachutist!

One serious point that Leonardo missed is that a parachute, contrary to common sense as it might appear at first sight, has to have a hole at the highest point to let the trapped air out! Without such a vent hole his parachute would oscillate from side to side as it descended and would move in a very unpredictable and potentially dangerous way. However, it is impressive that the size that Leonardo estimates as being suitable is very close to that of a modern parachute, vent or not, and it is interesting that Leonardo's drawing was not discovered until after the modern parachute had been reinvented.

Leonardo's drawing shows that his parachute is stiffened with a framework and that the apex is supported by a long pole which serves to maintain the shape. However, what happens to the man's head when he reaches the ground is too awful to contemplate!

Model 2. The Air Screw Helicopter

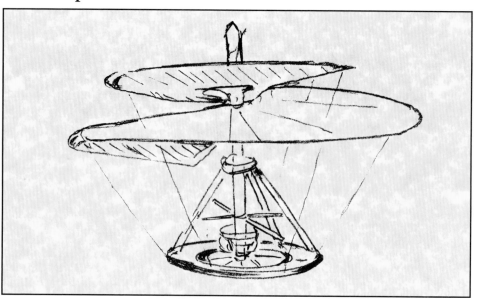

Helicopter

This splendid model seems to encompass everything that is wonderful about Leonardo's inventions and drawings. The idea is that as the helicopter rotates, it will screw itself up into the air rather as a metal woodscrew is drawn into the timber.

The power is provided by an operator or pilot who runs round on the base pushing on the handles which are attached to the main shaft of the helical screw propeller.

Leonardo states, *Let the outer extremity of the screw be of steel wire as thick as a cord, and from the circumference to the centre let it be eight arms lengths. I think, if this screw instrument is well made, that means made from linen of which the pores are stopped up with starch, and is turned rapidly, then the said screw will make its spiral in the air and it will rise high. Take the example of a wide and thin ruler whirled very rapidly in the air, you will see that your arm will be guided by the edge of the said flat surface.*

The framework of the above mentioned linen should be of long stout cane. You may make a small model of pasteboard, of which the axis is formed of fine steel wire, bent by force, and as it is released it will turn the screw.

It is true that if the helicopter could get up sufficient speed and did not weigh too much, that the screw would lift the machine into the air. It would of course be rotating furiously and the operator would soon begin to rotate with it unless Leonardo imagined that the platform somehow could remain stationary at the centre once he stopped running round and pushing. Perhaps he imagined the operator in a area of calm surrounded by the whirring sails looking out on the assembled crowd below. Of course our brave pilot should quickly learn to duck because his handle will still be rotating with the rest! In any case friction in the bearings would soon make everything rotate together. Even the modern helicopter with much better bearings, rotates as a whole if ever the secondary vertical rotor on the tail should fail.

Apart from all this, we know that things would not work at all as Leonardo imagined. The instant the machine leaves the ground there will be no frictional force to prevent the base rotating. So the base would be pushed by the man's feet and rotate in the opposite direction to the airscrew. He would not be able to keep the airscrew turning and the helicopter will descend to earth, assuming its enormous weight ever allowed it to leave the ground! It was not until Newton and his laws of motion that an idea like this was clearly formulated. 'Action and reaction are opposite and equal.'

Leonardo's Machines

Model 3. An Ornithopter in the Shape of a Boat

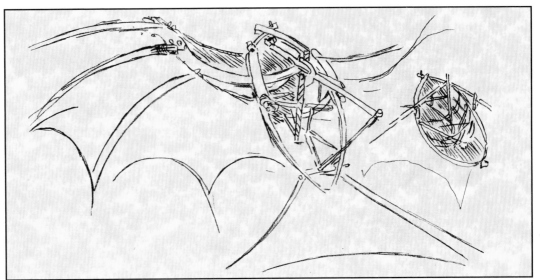

Flying Machine

This model is developed from one of Leonardo's earliest concepts for a flying machine and is based on a rowing boat fitted with wings instead of oars. The pilot uses a lever to operate the wings and at the same time turns a screw to apply more force to them.

Why did Leonardo choose a boat shaped fuselage? He knew that such a streamlined shape moves easily through the water and he might have seen the similarity between this shape and the shape of a bird's body as it flies, so he might have argued that this would slip through the air with little hindrance. In his later designs he abandons this idea because it gets in the way of incorporating more elaborate driving mechanisms and the weight penalty is huge.

The screw device in the middle of the ornithopter is interesting because it uses the principle of the 'inverted screw'. The top half of the screw which passes through a nut on the wing lever has a right-handed thread whereas the bottom half of the screw which passes through a nut fastened to the bottom of the boat has a left-handed thread. So when the operator turns the screw using the handle in the middle, it will move the wing lever twice as fast as a single screw would. This inverted screw would have required a separate operator with a corresponding increase in the weight to be carried.

Because we are modelling in card the screw operating mechanism has not been included. It is interesting that Leonardo soon realised that this idea had no future because he did not spend any more time developing it. He saw that using the powerful muscles in the legs would be a better bet and so turned his attention to other kinds of ornithopter.

The drawing also shows a tail plane rather like a bird's tail which is operated by a lever and which could be used to control the angle of elevation of the machine. The wings however, are based on the wings of a bat. It requires little imagination today to see that the machine would be very heavy whatever materials were used for its construction. However, aviation has taken the route of the fixed wing and the aerofoil section to create the lift necessary to keep the machine in the air. The line of development has come from the gliding action of birds rather than powered flapping. Early airplanes could be considered as gliders with engines and could glide a long way and land successfully should the engine fail. Modern ones, especially military planes where much higher performance and manoevrability is required, will scarcely glide at all and the only sensible response to engine failure is to use Leonardo's other invention, the parachute!

Model 4. The Wing Testing Machine

Wing Testing Machine

Leonardo states, *If you want to see a real test for the wings, make a wing from paper with a net and cane structure 20 arm lengths long and wide* (around 12 metres)*, and attach it on a plank which weighs 200 pounds; and apply, as shown, a sudden force. And if the 200 pound plank lifts itself before the wing descends then the trial can be considered successful; but be sure that the force is rapid and if the above effect is not obtained, waste no more time on it.*

Our model does not include the man but clearly shows the principle behind Leonardo's idea for a wing testing device. The beam holding the wing is pivoted on a bearing supported by a block of wood which is supposed to weigh 200 pounds. When the man pushes down on his lever the linkage will cause the wing to descend, the air resists the motion of the wing. This will result in an upward force on the bearing supporting the wing beam and if this force is great enough it will lift the bearing and the block of wood. You can simulate this in the model by holding your hand under the wing to stop its descent and the wood block will rise. The force of the air on a poor wing would not be sufficient to cause the block to be lifted.

Perhaps he was influenced by the way that an oar is used to row a boat. The sudden jerk when the blade enters the water and force is applied, creates an eddy and causes the blade to 'lock', so creating a pivot against which the boat can be levered. Because water is a fluid, this pivot moves during the stroke but none the less sufficient purchase can be obtained to drive the boat forward. Leonardo saw the wing as simply flapping up and down and his wing testing machine would certainly discover designs which were light and strong enough to survive this action and would 'lock' into the air efficiently. However, at the time he appears not to have paid enough attention to what happens on the return stroke. In a rowing boat the return stroke is made by either removing the oar from the water or by turning it sideways. Neither option was possible for his wing designs and more thought and research was needed.

Ideas for further study

There are many books, exhibitions and websites which are able to give far more space and detail to Leonardo's inventions than the small selection included in this collection. However, we hope that they have stimulated you to look further and to investigate and be astonished at what Leonardo was able to accomplish in his lifetime. Try if you can to look at a facsimile of a Codex in order to see the detail of the original drawings. Try also to understand exactly how he intended them to work and then see if you are able to find an error or a shortcoming. As we have seen throughout this selection of ideas and models, he was an outstanding genius but could still make mistakes, or was he protecting his ideas? He was after all, human!

How to Make Good Models

Glueing

To get the best results you need a glue which sets quickly but not instantly and which does not leave dirty marks. We particularly recommend a petroleum based glue such as 'UHU All Purpose', especially the 'gel' version. Both 'BOSTIK Clear', white latex adhesives such as 'Copydex' and PVA also give good results. When applying the glue it is a good idea to use a cocktail stick or small piece of card as a spatula. When handling small parts a pair of fine tweezers can be useful.

Sticks and Threads

The 2.5mm diameter stick-holes are designed to take cocktail sticks. It is a good idea to pierce these holes with the point of a pair of compasses and then widen them out to an exact fit using the sharp end of a cocktail stick. Pierce the thread holes with a point before glueing and stiffen the end of the thread with a smear of glue so it can easily be passed through the holes to complete the model.

Scoring

Scoring is very important if you are to make accurate models. It makes the paper fold cleanly and accurately along the line you want. Use a ball point pen which has run out of ink and rule firmly along the fold lines. Experienced model makers may use a craft knife, but it needs care not to cut right through the paper.

Hill and Valley Folds

1. Cut out the model of your choice from the book, keeping well away from the outline.
2. Score along all fold lines and then cut out the individual pieces precisely.
3. Carefully fold along the fold lines to make a hill or a valley fold according to whether the line is dashed or dotted.

A solid cut line

On dashed lines fold away from you to give a hill fold.

On dotted lines fold towards you to give a valley fold.

4. Crease firmly.
   Do not start to glue until all the creases are done.
5. Then follow the instructions given on the next few pages. Noting that the grey areas indicate where one piece glues over another. All, except those on the parachutist's face are covered when the models are complete.

How to Make Leonardo's Parachute

Stage 1: The Tent and Post (2 pieces required)

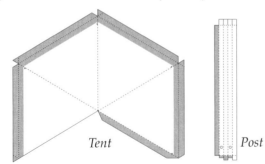

Tent   Post

1. Make the Tent and glue up the flaps to give it strong edges.
2. Make the Post and seal the end near the stick holes.
3. Glue the top of the Post inside the Tent's apex.

Stage 2: The Parachute Man (2 pieces, 25mm stick and thread required)

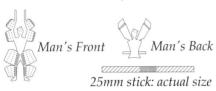

Man's Front   Man's Back

25mm stick: actual size

1. Make the arms by glueing the flaps on the Man's Front backwards to form triangular prisms.
2. Make the legs by glueing the flaps on the Man's Front forwards.
3. Glue the Man's Back to the Man's Front, matching his hat.
4. Bring the two halves of the doublet forwards and glue the belt in position.

Stage 3: Assembling the Parachute

1. Place the stick through the stick holes and attach the Man on either side of the Post.
2. Add the threads between the eye holes on the Tent and the stick.

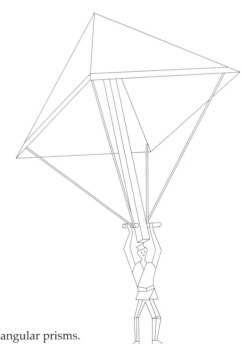

Completed Model

How to Make Leonardo's Helicopter

Stage 1: The Barrel and Platform (14 pieces required)

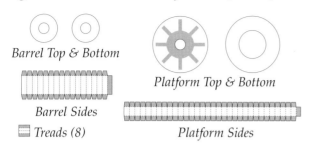

Barrel Top & Bottom

Platform Top & Bottom

Barrel Sides

Treads (8)

Platform Sides

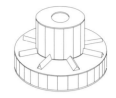

Barrel and Treads on the Platform

1. Make the Barrel and glue it on to the Platform Top, matching the cut out centres.
2. Then make the 8 triangular treads and glue in position before completing the Platform.

Stage 2: Assembling the Central Structure (14 pieces required)

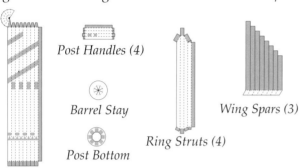

Post Handles (4)

Barrel Stay

Wing Spars (3)

Ring Struts (4)

Post Bottom

Central Post

Post Bottom

Barrel Stay

Post Handles

Ring Struts

1. Glue the Central Post with its capped top.
2. Put the bottom flaps of post through the Post Bottom and glue the flaps outwards on the under side.
3. Place the Platform/Barrel on the Central Post and glue the Barrel Stay in position.
   (The Barrel/Platform should rotate freely between the Post Bottom and the Barrel Stay.)
4. Make the Handles glueing them in position, as shown, on the Central Post.
5. Make the Ring Struts and glue them in position.
6. Next glue the spiralling Wing Spars in position with the smaller spars towards the top.
   (Note that the Spars lie above their glue flaps.)

Stage 3: Completing the Helicopter (22 pieces and thread required)

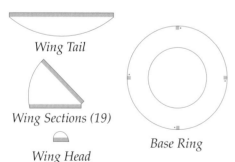

Wing Tail

Wing Sections (19)

Wing Head

Base Ring

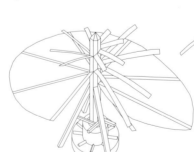

Glueing the Wing

Wing Spars

1. Glue the Wing under the Wing Spars, Section by Section.
   (Start with the Wing Tail at the bottom, glueing each section close to the Central Post.)
2. If necessary, trim the ends of each Wing Spar back to the Wing.
3. Glue the Base Ring to the bottom of the Ring Struts.
4. Add the four threads between the eye holes on the Wing and on the Ring.

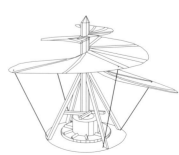

Completed Model

How to Make Leonardo's Ornithopter

Stage 1: The Boat *(6 pieces required)*

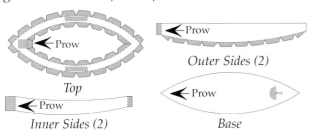

Top

Outer Sides (2)

Inner Sides (2)

Base

Boat

1. Glue the Inner Sides together with the double thickness at the Prow, then glue to the Top, starting at the prow end.
2. Glue the Outer Sides together at the wider end, then glue to the Top starting with this wide joint at the prow. (Working back towards the stern takes up any inaccuracies in cutting and folding.)
3. Glue the Base in position, with the printed side on the outside and the stick slot towards the stern.

Stage 2: The Wings and Fin *(12 pieces and 50mm stick required)*

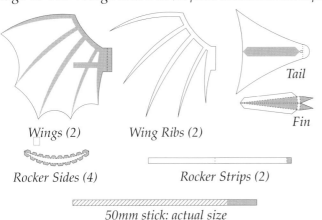

Wings (2)

Wing Ribs (2)

Tail

Fin

Wing/ Wing Rib

Making the Rockers

Rocker Sides (4)

Rocker Strips (2)

50mm stick: actual size

Wing with Rocker

Tail and Fin

1. Glue the Wing Ribs in position on the Wings, then fold each hinge downwards below the Wings.
2. Make the 2 Rockers and attach them to the Wing/Wing Rib pieces.
3. Glue main grey area of Fin back to back, then glue the stick in position. (The stick diagram above shows where to glue.)
4. Glue the pointed ends of the Fin together, then glue the Fin to the Tail putting the stick through the round stick hole.

Stage 4: Assembling the Ornithopter *(3 pieces and 2 65mm sticks required)*

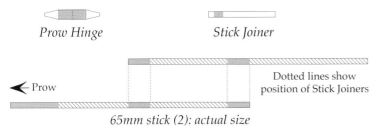

Prow Hinge

Stick Joiner

Prow

Dotted lines show position of Stick Joiners

65mm stick (2): actual size

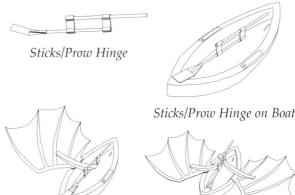

Sticks/Prow Hinge

Sticks/Prow Hinge on Boat

1. Join the 2 65mm sticks together, as shown, using the 2 Stick Joiners. (This makes a rigid structure.)
2. Glue the rectangular section of the Prow Hinge back against itself.
3. Wrap and glue the narrow ends of the Prow Hinge around the end of one of the joined sticks.
4. Glue the joined sticks in position on the Boat using the Prow Hinge.
5. Attach the Wings to the Top. (The ends of the Rockers are placed through the joined sticks.)
6. Put the Fin stick through the slot in the Base and then glue the Tail to the bottom of the Base making a hinge.

Wings with Rockers through Sticks

Completed Model

23

How to Make Leonardo's Wing Tester

Stage 1: The Base with its Trusses (15 pieces required)

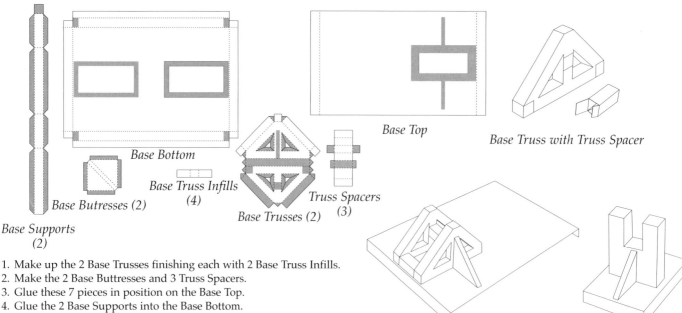

Base Bottom

Base Top

Base Truss with Truss Spacer

Base Truss Infills (4)

Base Butresses (2)

Base Trusses (2)

Truss Spacers (3)

Base Supports (2)

Base Top complete

Block complete

1. Make up the 2 Base Trusses finishing each with 2 Base Truss Infills.
2. Make the 2 Base Buttresses and 3 Truss Spacers.
3. Glue these 7 pieces in position on the Base Top.
4. Glue the 2 Base Supports into the Base Bottom.
5. Complete the Base by glueing the Base Bottom and Base Top together.

Stage 2: The Block with its Truss (7 pieces required)

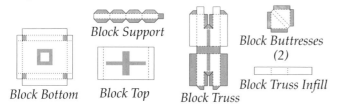

Block Support

Block Buttresses (2)

Block Bottom

Block Top

Block Truss

Block Truss Infill

1. Make up the Block Truss complete with Block Truss Infill.
2. Make the 2 Block Buttresses.
3. Glue these 3 pieces in position on the Block Top.
4. Glue the Block Support into the Block Bottom.
5. Complete by glueing the Block Bottom to Block Top.

Stage 3: The Base and Block Shafts
(13 pieces, 40mm stick, 30mm stick and 2 25mm sticks required)

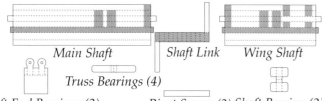

Main Shaft

Shaft Link

Wing Shaft

Truss Bearings (4)

Shaft End Bearings (2)

Pivot Spacer (2) Shaft Bearing (2)

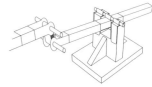

Shaft End Bearing in place

Shaft Link zig zags glued

Shafts linked

Hinge on Block Truss

1. Make the Main Shaft and Wing Shaft, these each have a diagonal strip to give support.
2. Glue the 2 Shaft End Bearings to the marked ends of the Shafts.
3. Glue the Shaft Link zig-zags. Then join the shafts using the 2 25mm sticks and the Shaft Link, forming a free moving joint.
4. Attach the 30mm stick firmly to the Block Truss using 2 Truss Bearings.
5. Then using a Shaft Bearing join the Wing Shaft to the Block Truss, creating a free moving hinge.
6. Similarly attach the Main Shaft to the Base Truss. Wrap and glue the 2 Pivot Spacers on to the stick to keep the Main Shaft central.

Stage 4: The Wing (4 pieces required)

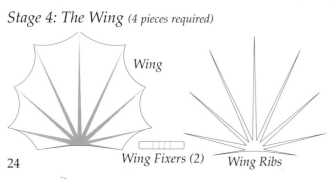

Wing

Wing Fixers (2)

Wing Ribs

1. Glue the Wing Ribs to the Wing.
2. Use the 2 Wing Fixers to glue the Wing to the Wing Shaft. (The base of the Wing lies up against the Shaft Bearing.)

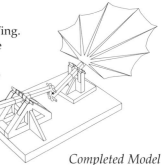

Completed Model

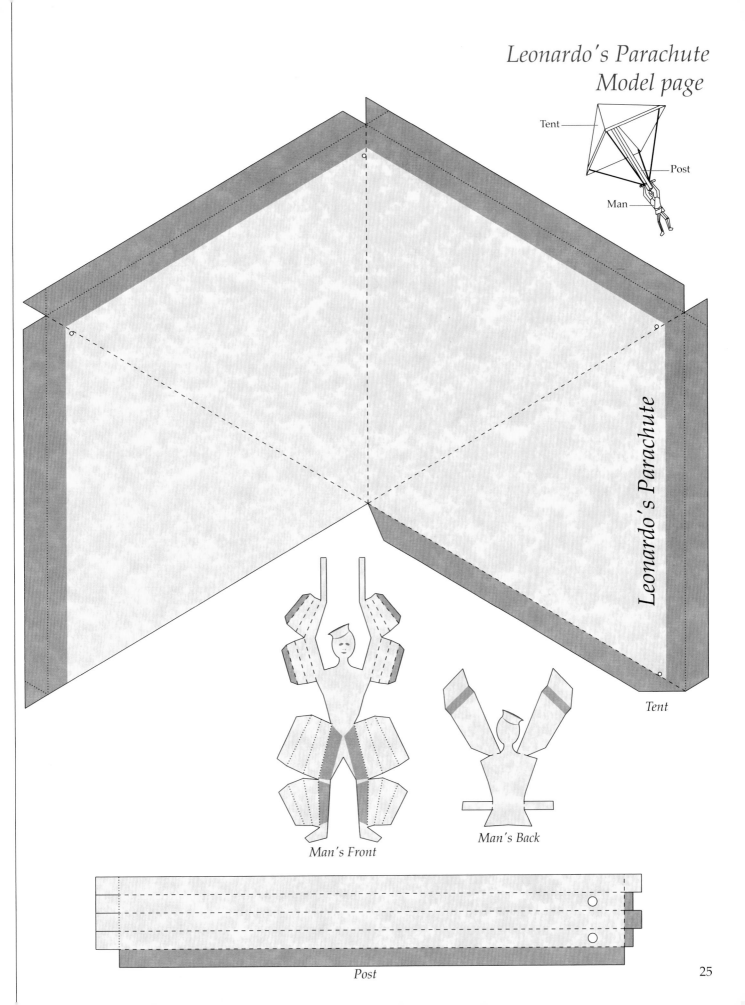

Tent

Post

Man

Leonardo's Parachute

Tent

Man's Front

Man's Back

Post

25

Wing section 15

Wing section 14

Wing section 13

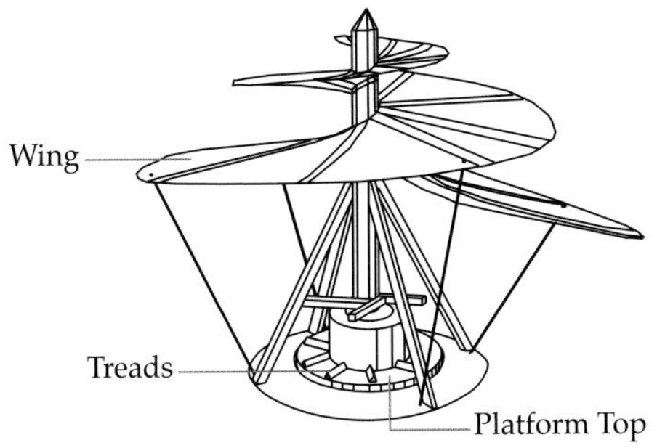

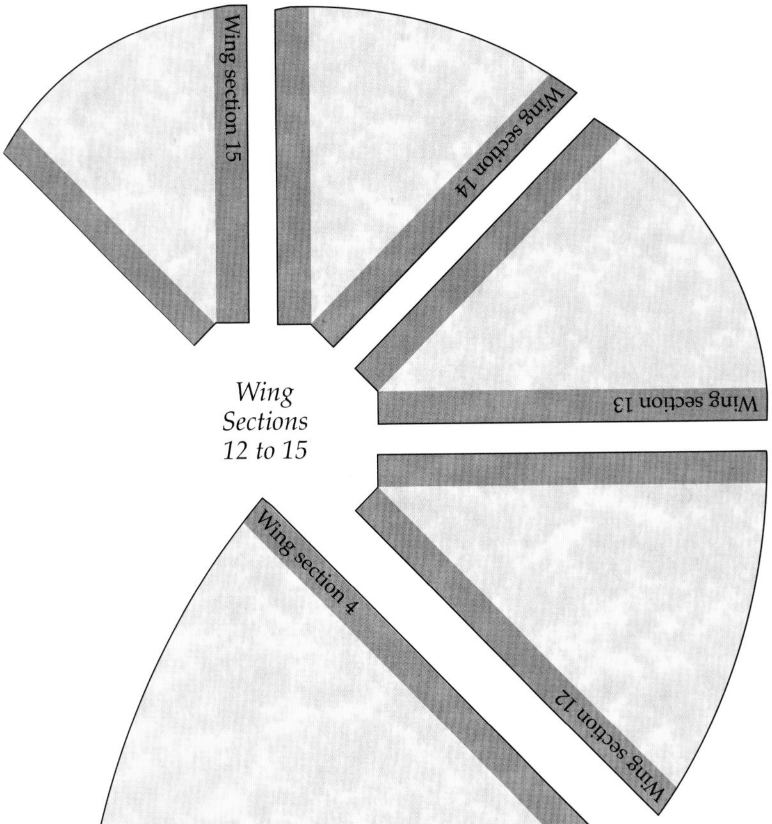

*Wing
Sections
12 to 15*

Wing section 12

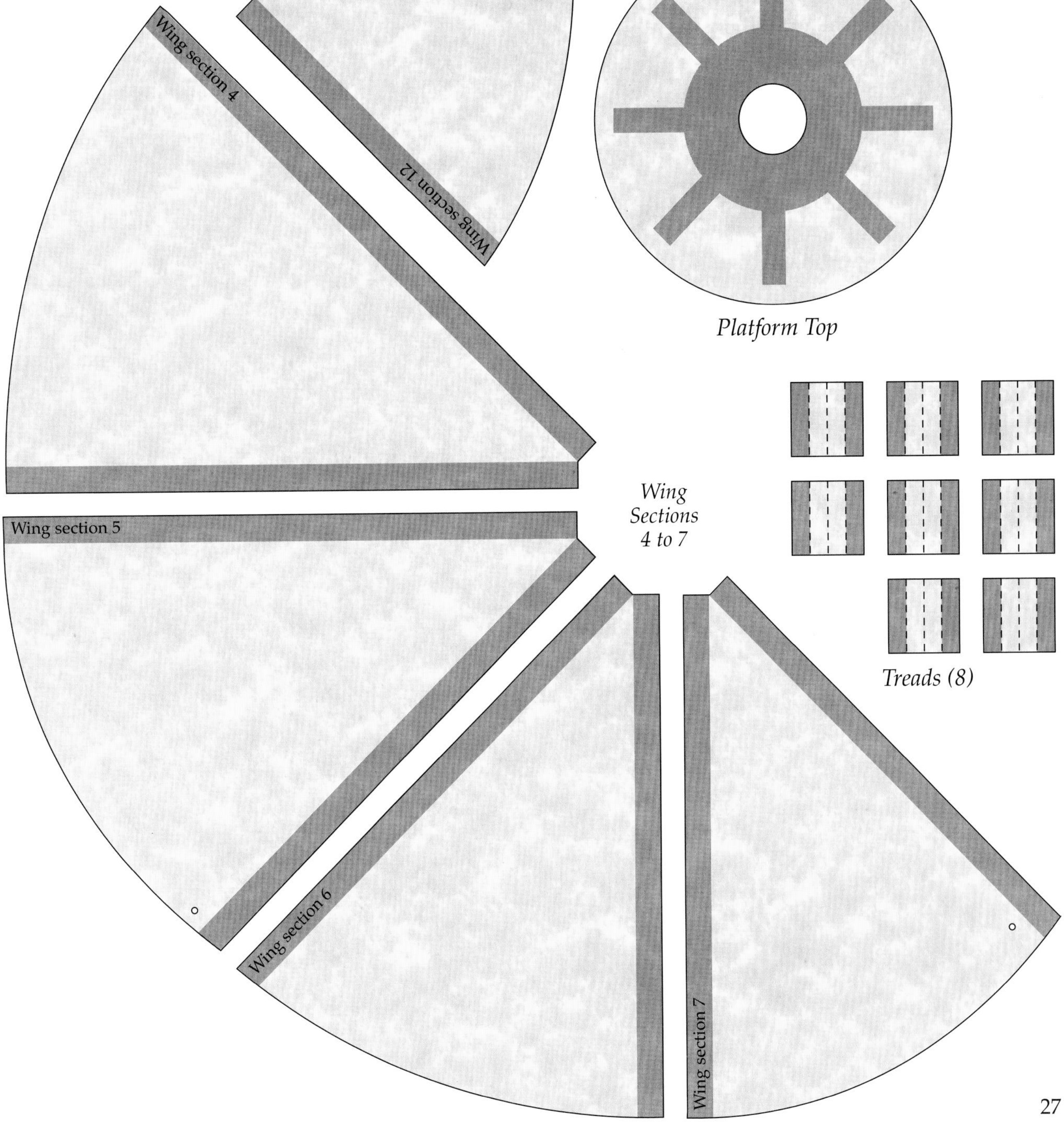

Wing section 4

Wing section 5

Wing section 6

*Wing
Sections
4 to 7*

Wing section 7

Wing

Treads

Platform Top

Platform Top

Treads (8)

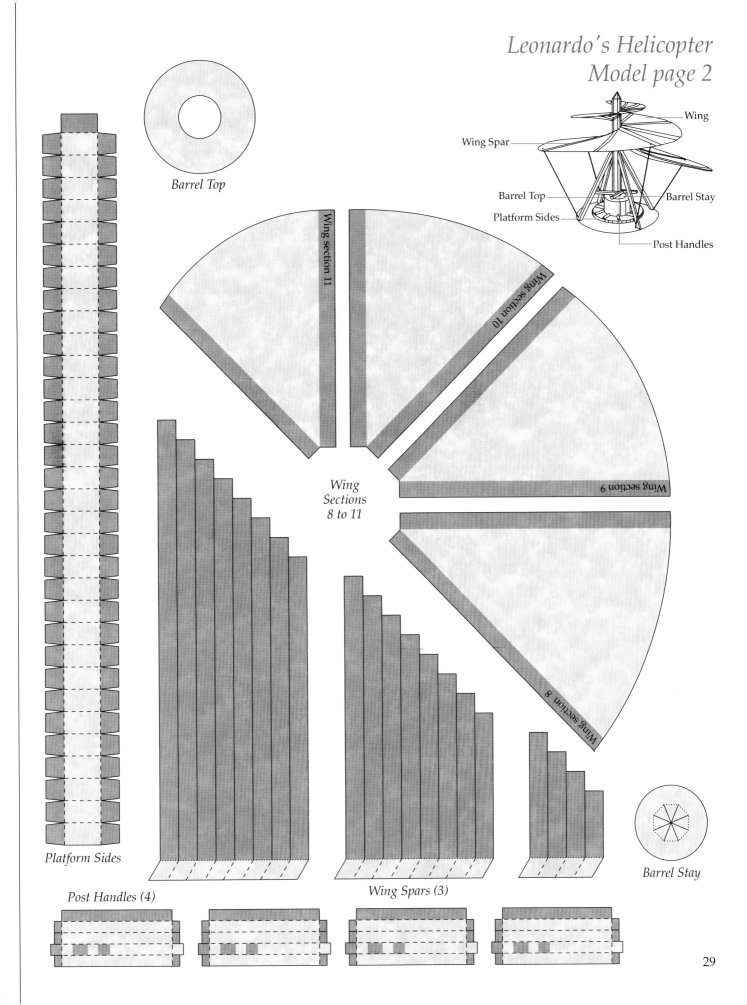

Wing

Wing Spar

Barrel Top

Barrel Stay

Platform Sides

Post Handles

Barrel Top

Wing section 11

Wing section 10

Wing section 9

Wing section 8

Wing Sections 8 to 11

Platform Sides

Barrel Stay

Wing Spars (3)

Post Handles (4)

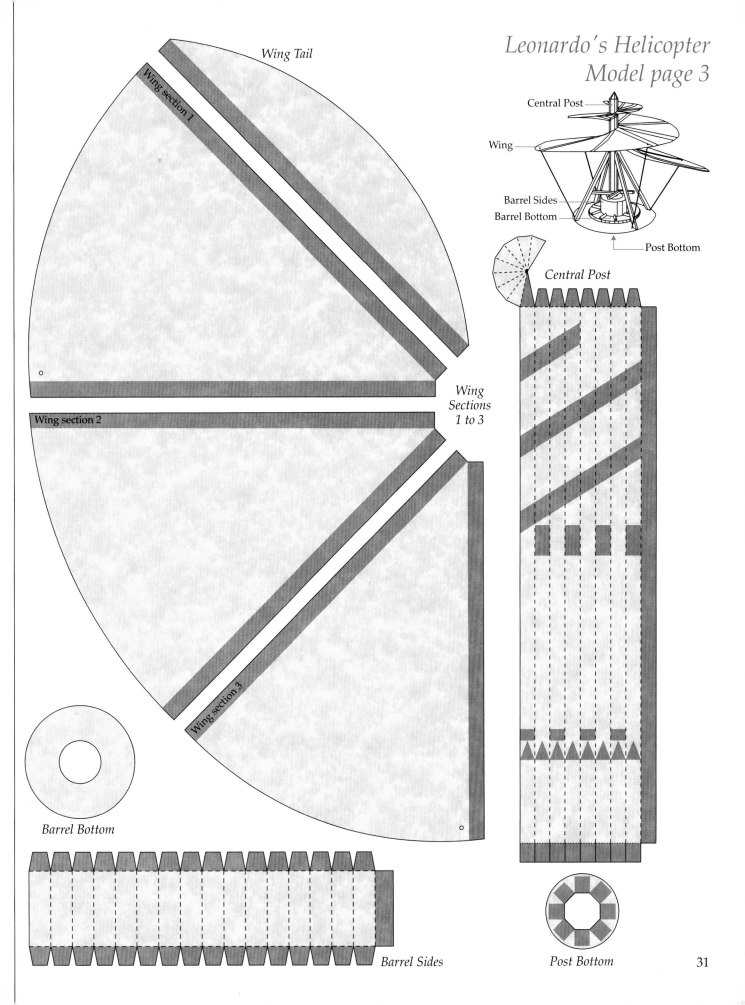

Wing Tail

Wing section 1

Wing section 2

Wing section 3

Wing Sections 1 to 3

Central Post

Wing

Barrel Sides

Barrel Bottom

Post Bottom

Central Post

Barrel Bottom

Barrel Sides

Post Bottom

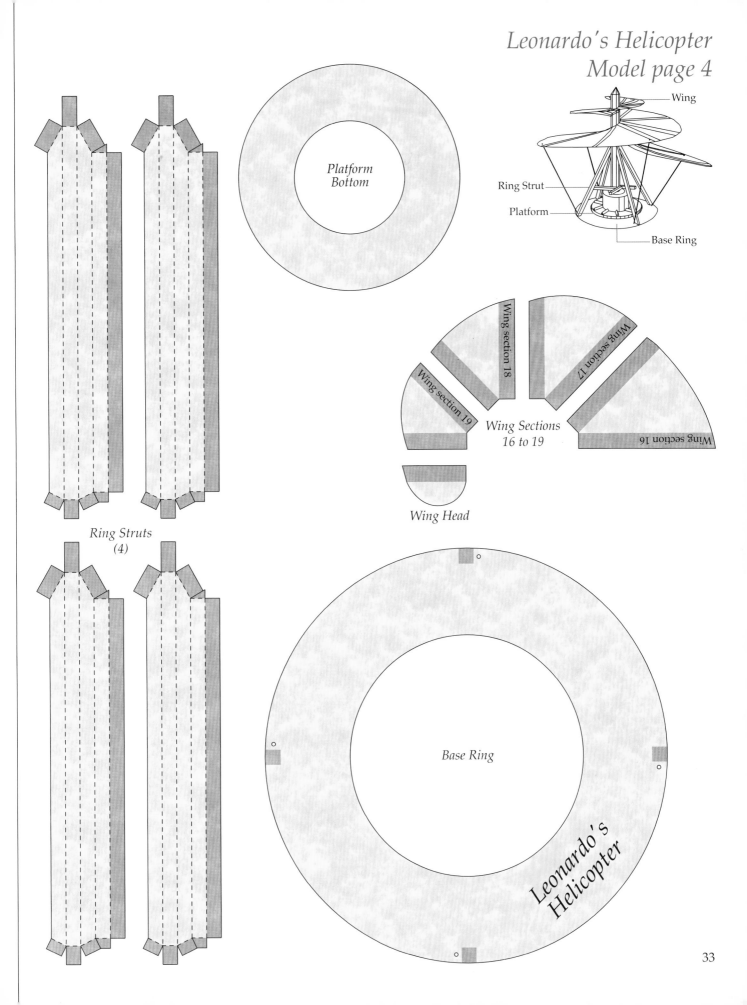

Wing

Ring Strut

Platform

Base Ring

Platform
Bottom

Wing section 19

Wing section 18

Wing section 17

Wing section 16

*Wing Sections
16 to 19*

Wing Head

*Ring Struts
(4)*

Base Ring

Leonardo's
Helicopter

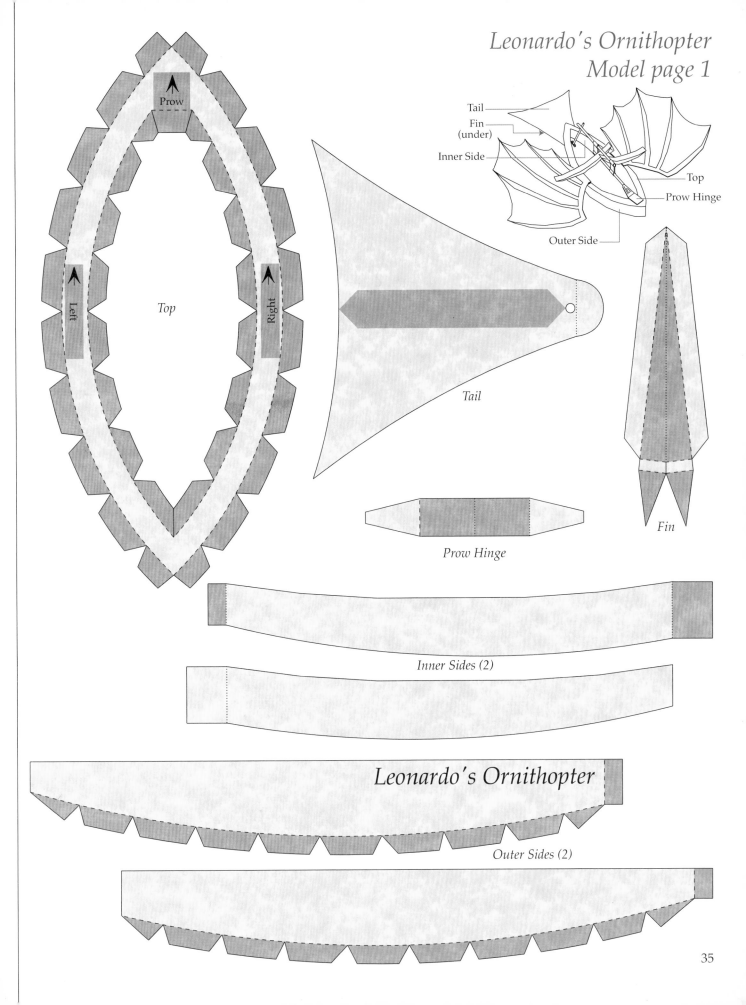

Prow

Left

Right

Top

Tail
Fin
(under)
Inner Side
Top
Prow Hinge
Outer Side

Tail

Fin

Prow Hinge

Inner Sides (2)

Leonardo's Ornithopter

Outer Sides (2)

35

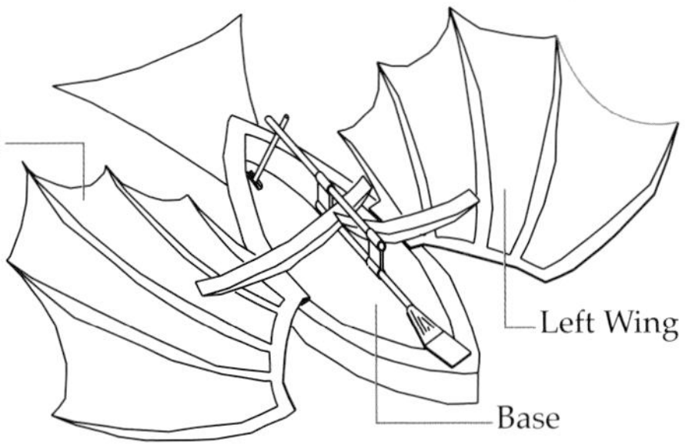

Right Wing

Left Wing

Base

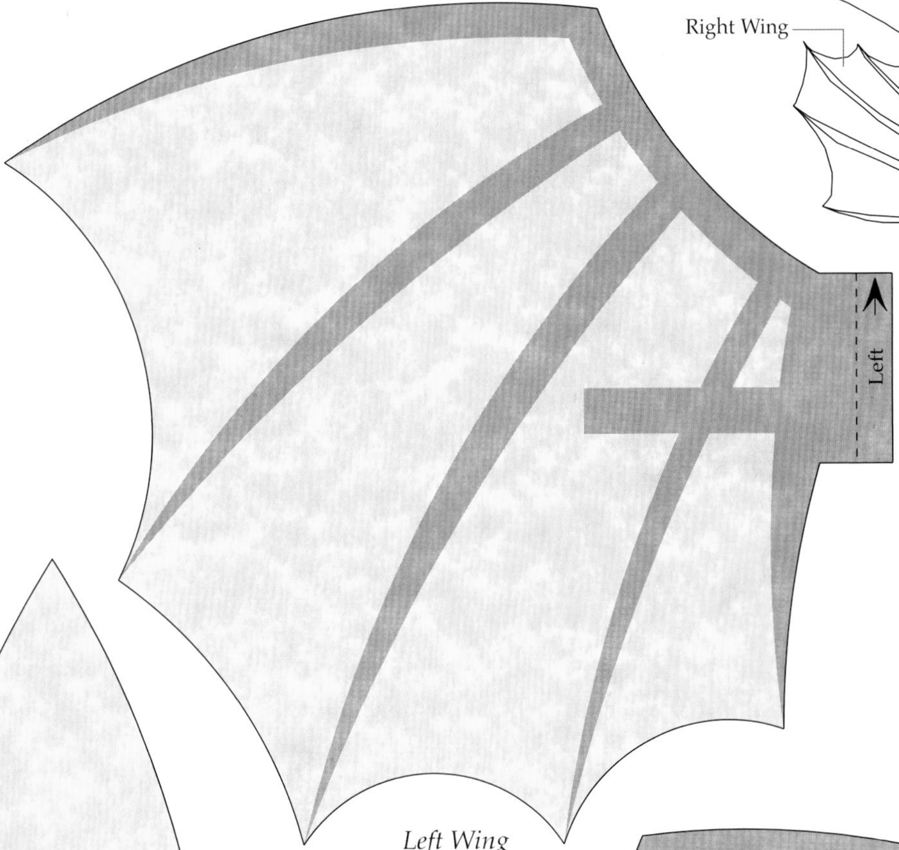

Left

Left Wing

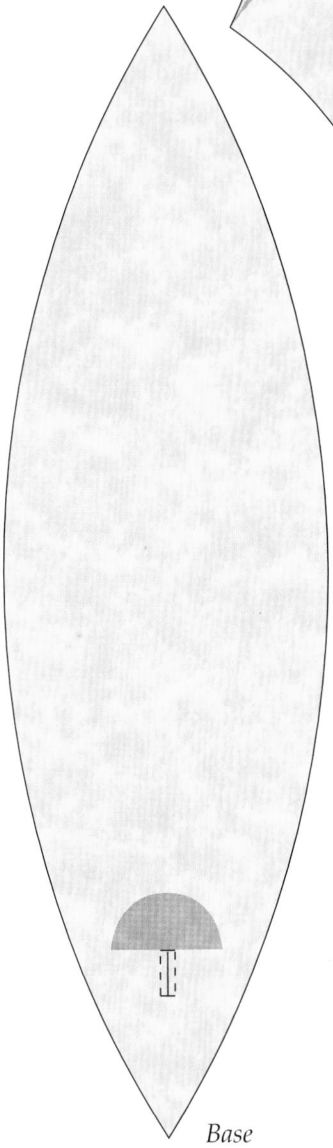

Right

Base

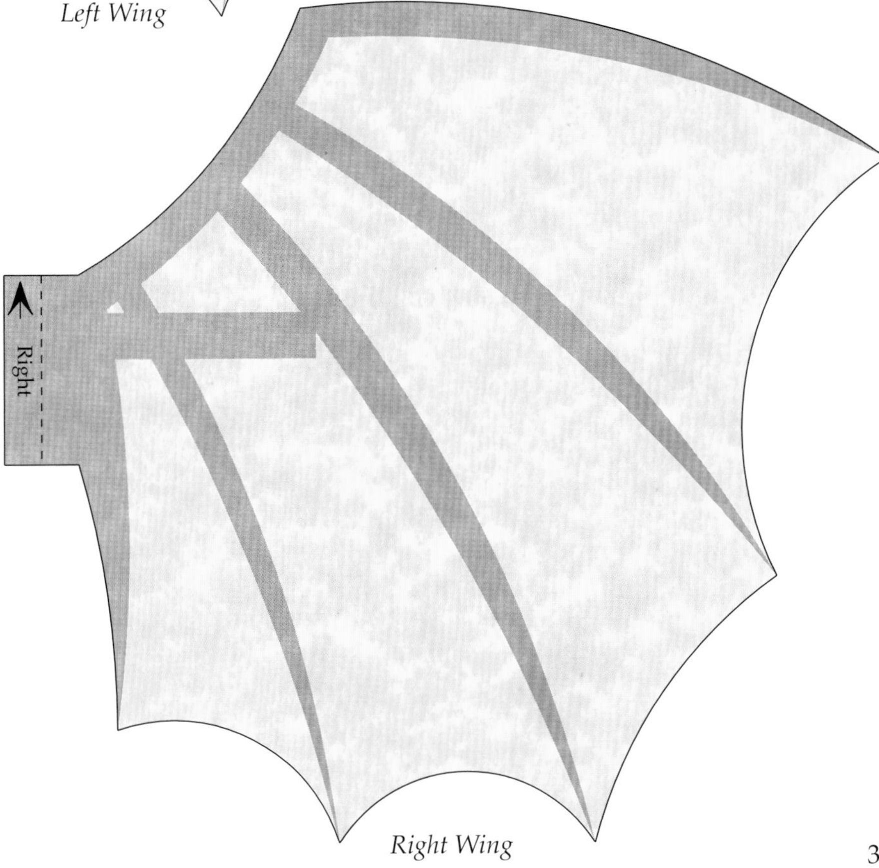

Right Wing

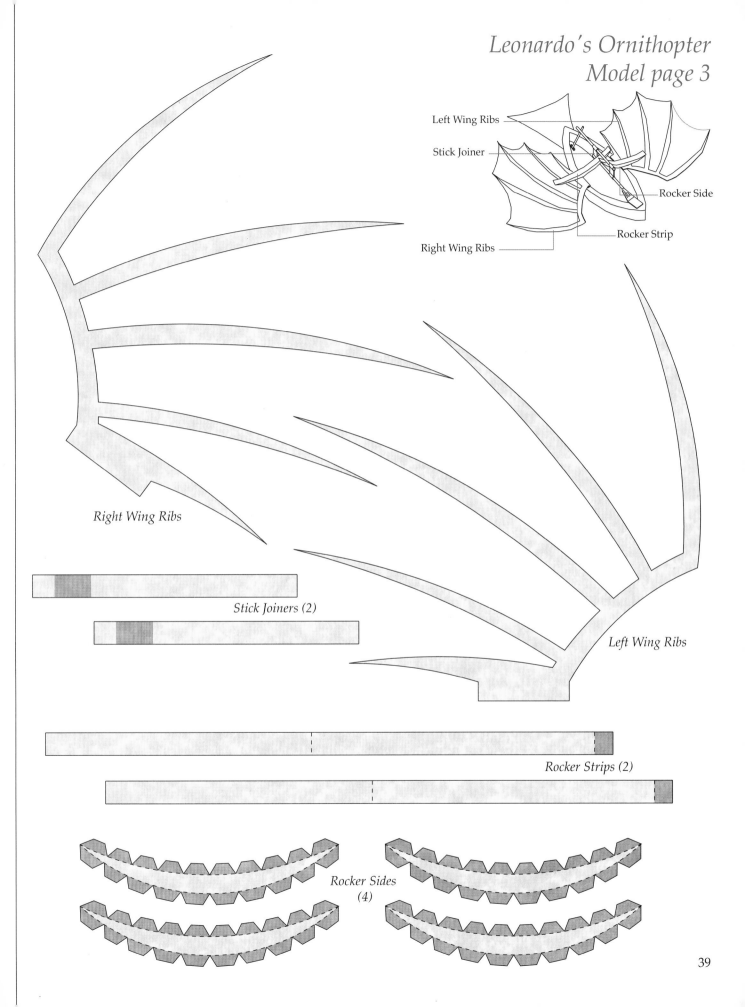

Left Wing Ribs

Stick Joiner

Rocker Side

Rocker Strip

Right Wing Ribs

Right Wing Ribs

Stick Joiners (2)

Left Wing Ribs

Rocker Strips (2)

Rocker Sides
(4)

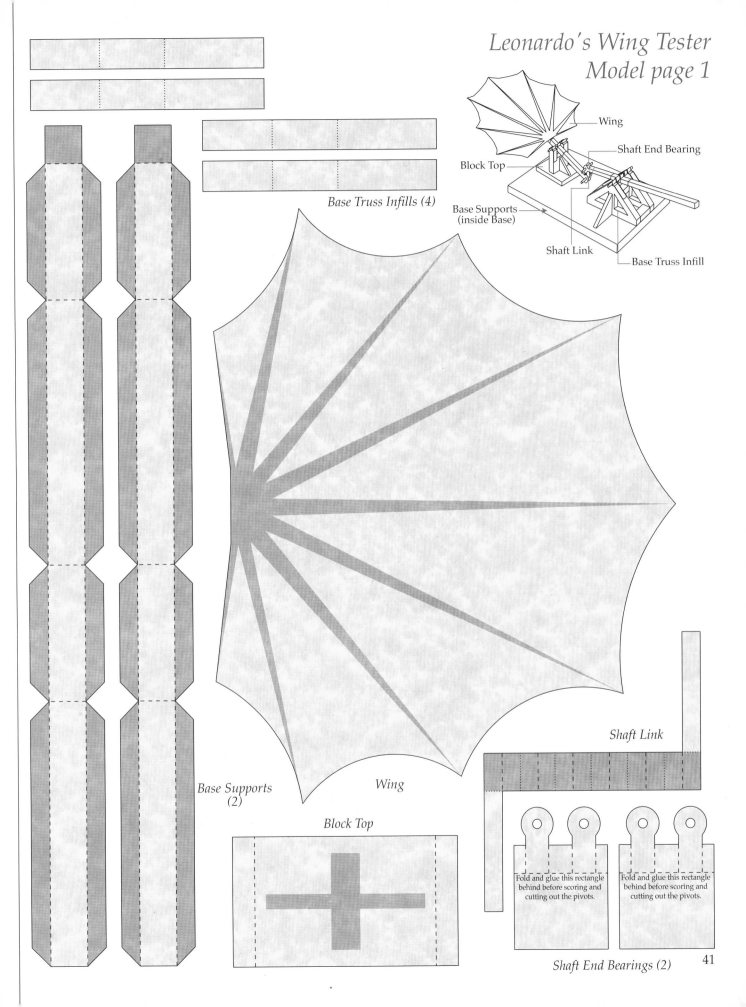

Wing

Shaft End Bearing

Block Top

Base Supports
(inside Base)

Shaft Link

Base Truss Infill

Base Truss Infills (4)

*Base Supports
(2)*

Wing

Shaft Link

Block Top

Fold and glue this rectangle
behind before scoring and
cutting out the pivots.

Fold and glue this rectangle
behind before scoring and
cutting out the pivots.

Shaft End Bearings (2)

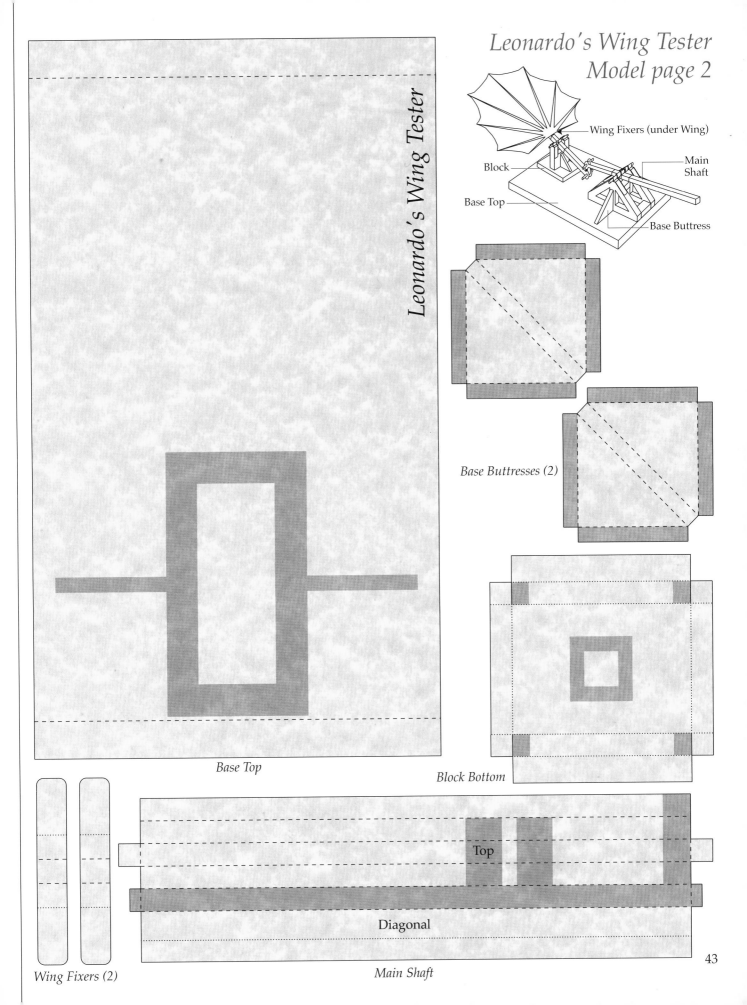

Leonardo's Wing Tester

Wing Fixers (under Wing)

Block

Main Shaft

Base Top

Base Buttress

Base Buttresses (2)

Base Top

Block Bottom

Top

Diagonal

Wing Fixers (2)

Main Shaft

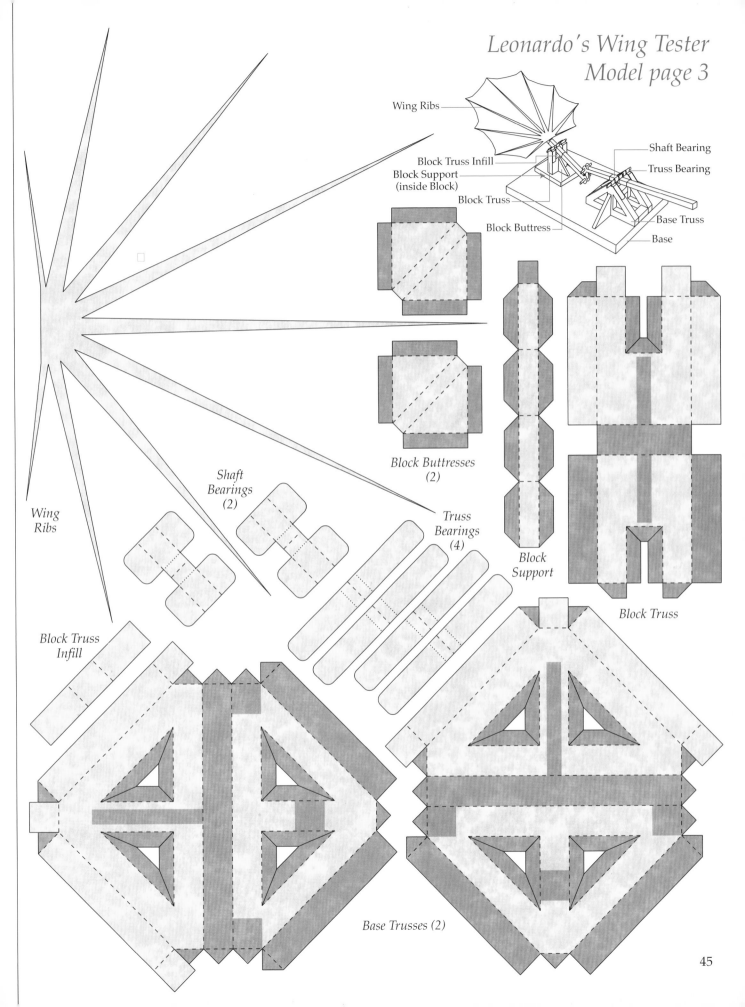

Wing Ribs

Block Truss Infill

Block Support
(inside Block)

Block Truss

Block Buttress

Shaft Bearing

Truss Bearing

Base Truss

Base

Block Buttresses
(2)

Shaft
Bearings
(2)

Truss
Bearings
(4)

Block
Support

Block Truss

Wing
Ribs

Block Truss
Infill

Base Trusses (2)

45

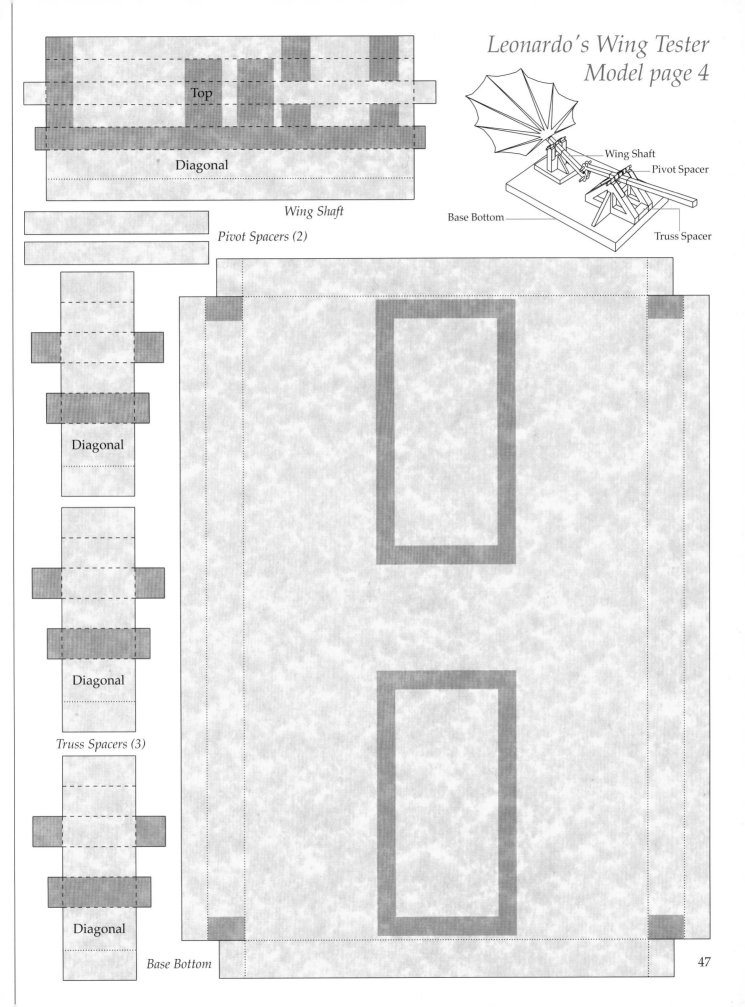

Top

Diagonal

Wing Shaft

Wing Shaft

Pivot Spacer

Base Bottom

Truss Spacer

Pivot Spacers (2)

Diagonal

Diagonal

Truss Spacers (3)

Diagonal

Base Bottom